## CONTENTS: | TABLE OF CONTENTS

## CONTENTS: | WHAT IS THIS BOOK ABOUT?

**STEM** is an acronym for Science, Technology, Engineering, and Mathematics. Recent shifts in education have favored these subjects, primarily because we have a shortage of workforce in these particular areas, which is really quite sad for a number of reasons. These can be some of the most interesting things to study in school, provided they are taught in a fun, interesting, and hands-on fashion. They also lead to some of the best-paying technical jobs, too!

All of the labs within promote these 4 fields. You will find a strong emphasis on designing a project, testing it, measuring the results, and reflecting upon what worked and did not work. The projects are also labeled at the bottom with a series of categorical tags, so you can find similar projects to work on, allowing students to build upon prior knowledge gained in doing these projects! Of course, you can do them in any order you wish, but it can be fun to do a set of similar projects.

Since this is an educational volume, developed in my years teaching science in public schools, grading rubrics for each assignment are provided. There are some general suggestions and guidelines for each project, but it has deliberately been left without too much detail to allow the projects to be adapted to your classroom's individual needs.

## CONTENTS: | COPYRIGHT INFORMATION

Curriculum Knowledge Center
CEHHS
University of Michigan Dearborn

## CONTENTS: Alphabetical Mission Listing - Page 1

## CONTENTS: Alphabetical Mission Listing - Page 2

## CONTENTS: Mission Categories Listing

- Accuracy
- Arches
- Balloons
- Boats
- Bridges
- Buoyancy
- Cable Cars
- Cantilevers
- Capacity
- Cardboard Tubes
- Cars
- Clay
- Coins
- Crashes
- Dead Lift
- Distance
- Fliers
- Foil
- Gears
- Height
- Length
- Levers
- Marbles
- Marshmallows
- Materials Strength

- Paper
- Pasta
- Ping Pong Balls
- Pipe Cleaners
- Plastic Straws
- Plastic Wrap
- Popsicle Sticks
- Pulleys
- Rotation
- Rubber Bands
- Scavengers
- Sorting
- Speed
- String
- Throwers
- Time
- Tissue
- Toothpicks
- Towers
- Tracks
- Trampolines
- Water
- Wax Paper
- Weight
- Wind

# CONTENTS: Categorical Mission Listing - Page 1

# CONTENTS: Categorical Mission Listing - Page 2

## CONTENTS: Categorical Mission Listing - Page 3

# CONTENTS: Categorical Mission Listing - Page 4

## CONTENTS: Categorical Mission Listing - Page 5

## CONTENTS: 50 MORE STEM LABS

Each of the new science labs in **50 MORE STEM LABS** has the following:

- A snappy **Title**

- A **Brief Description** of the task to be completed

- General **Mission Rules**, suggestions, limitations, and requirements of the task

- **Grading Rubrics** for a Quiz and a Test Grade

- A small **note space** for any changes or adaptations required

- **Category Tags** at the bottom to help you find similar projects

## MISSION: Bayou Fanboats

**BRIEF:** You and your team have been selected to make balloon-powered boat that goes as fast and as far as possible.

### MISSION RULES:

1. You will design a boat that is balloon-powered and attached to a string. The rocket ship must slide along the string to cross the body of water.

2. Your boat must be built from a single balloon, a straw, tape or glue, notecards, and other teacher-approved materials.

3. You will work with a single partner. Teams may not be of more than 2 people.

4. The straw will be used to slide along the line that is strung horizontally above a body of water. Both ends will be secured during tests. Inflated balloons will be attached to the balloon boat. Letting air from the balloons should propel the project across the water as fast as possible.

5. If your boat does not clear the distance with one use of the balloon, your teacher has the option of measuring the distance traveled or keeping the clock running while you refill the balloon as many times as needed to clear the course.

*TEACHER'S NOTES: You will need some sort of body of water. A set of 1-3 rain gutters coupled together with caps at the end will hold water. Then you can run the string/fishing line parallel to the surface of the water.*

### QUIZ GRADE:

Create a blueprint designs for your ideas

- Sketch 25%

- Sketch is labeled 25%

- Explanation of strategies 25%

- Conclusions and reflections based on your results 25%

### TEST GRADE:

Your completed design and the results of the test.

- Project Completed = 50%

50% of your grade depends on how fast your project is compared to others.

- Other places = +10-20%

- Third Place = +30%

- Second Place = +40%

- Fastest Boat = +50%

### NOTES:

**CATEGORIES:** Balloons, Boats, Buoyancy, Distance, Speed, Water

## MISSION: Ben Franklin's Kites

**BRIEF:** You and your team have been selected to make a kite using tin foil!

### MISSION RULES:

1. You will design a kite.

2. You will work with one or two partners. Teams may not be of more than 3 people.

3. The kite must be built using foil as your primary building material. It is up to your teacher how much and what kind of other materials are allowed.

4. Teams are encouraged to make non-traditional designs. Diamond-shaped kites will not get full points! Kite research is strongly recommended.

*TEACHER'S NOTES: Obviously, this should not be done in poor weather. :)*

### QUIZ GRADE:

Create a blueprint design for your ideas

- Sketch 25%

- Sketch is labeled 25%

- Explanation of strategies 25%

- Conclusions and reflections based on your results 25%

### TEST GRADE:

Your completed design and the results of the test.

- Project Completed = 50% (25% for diamond design)

- Project Gets airborne for 10+ seconds = 25%

- The remaining 25% of your points come from how long your project remains airborne compared to other teams' kites.

### NOTES:

**CATEGORIES:** Fliers, Foil, Height, String, Wind

## MISSION: Cable Cars I - Weight

**BRIEF:** You and your team have been selected to make a cable car that can hold as much weight as possible.

### MISSION RULES:

1. You will design a cable car that is attached to a string with a sliding plastic straw. The cable car must attach to the string, but it does not have to slide.

2. Your cable car must be built from a plastic straw, tape or glue, notecards, and other teacher-approved materials.

3. You will work with a one or two partners. Teams may not be of more than 3 people.

4. Your device must have a cup or receiver for weights to be placed in. If the cable car breaks or if your cup is full, no more weight will be added and the test will be over.

*TEACHER'S NOTES: Suggested weights are pennies, paper clips, sand, or graduated weights.*

*For the line, heavy fishing line is suggested, anchored and tied on one end. The other end should be tied and retied for each test. Some sort of eye-bolt or carabiner might help. You might need to put all the straws on the line FIRST, and then tie it. Cable cars would then be designed in place.*

### QUIZ GRADE:

Create a blueprint designs for your ideas

• Sketch 25%

• Sketch is labeled 25%

• Explanation of strategies 25%

• Conclusions and reflections based on your results 25%

### TEST GRADE:

Your completed design and the results of the test.

• Project Completed = 50%

50% of your grade depends on how much weight your project holds compared to others.

• Other places = +10-20%

• Third Place = +30%

• Second Place = +40%

• Holds the Most Weight = +50%

### NOTES:

### CATEGORIES: Cable Cars, Materials Strength, Scavengers, String, Weight

# MISSION: Cable Cars II - Water

**BRIEF:** You and your team have been selected to make a cable car that can hold as much water as possible.

## MISSION RULES:

1. You will design a cable car that is attached to a string with a sliding plastic straw. The cable car must attach to the string, but it does not have to slide.

2. Your cable car must be built from a plastic straw, plastic, styrofoam, foil, tape or glue, notecards, and other teacher-approved materials.

3. You will work with a one or two partners. Teams may not be of more than 3 people.

4. Your device must have a cup or receiver for water to be placed in. If the cable car breaks, leaks, or your cup is full, no more water will be added and the test will be over.

*TEACHER'S NOTES: Keep a bucket under the test projects or test outside.*

*For the line, heavy fishing line is suggested, anchored and tied on one end. The other end should be tied and retied for each test. Some sort of eye-bolt or carabiner might help. You might need to put all the straws on the line FIRST, and then tie it. Cable cars would then be designed in place.*

## QUIZ GRADE:

Create a blueprint designs for your ideas

- Sketch 25%

- Sketch is labeled 25%

- Explanation of strategies 25%

- Conclusions and reflections based on your results 25%

## TEST GRADE:

Your completed design and the results of the test.

- Project Completed = 50%

50% of your grade depends on how much water your project holds compared to others.

- Other places = +10-20%

- Third Place = +30%

- Second Place = +40%

- Holds the Most Water = +50%

## NOTES:

**CATEGORIES:** Cable Cars, Capacity, Materials Strength, Scavengers, String, Water

# MISSION: Cantilever Catastrophe I - Paper

**BRIEF:** You and your team have been selected to make the longest cantilever possible from just tape and paper.

## MISSION RULES:

1. Your cantilever must between 2 and 6 inches wide, and 2 to 12 inches tall, but as long as possible. If you are outside these height and width measurements by more than 1/2 inch, you will be penalized.

2. You will work with one or two partners. Teams may not be of more than 3 people.

3. You must only use paper and glue, and tape for your project.

4. The cantilever may be taped to the desk or table on one end.

5. A beam of sorts will project away from the table horizontally as far as possible. The beam distance will only be counted as far as it goes out from the table edge, provided it does not drop more than 2 inches below the table's surface. The distance will only count up to the point in the beam before it descends the acceptable level.

*TEACHER'S NOTES: For a more difficult task, add a small weight to the end of the beam, like a penny or large paper clip.*

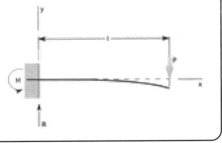

## QUIZ GRADE:

A research paper on cantilevers.

- 2-4 pictures of cantilevers 25%

- A labeled sketch and detailed concept idea based on your cantilevers pictures 50%

- Conclusions and reflections based on your results 25%

## TEST GRADE:

Your completed design and the results of the test.

- Project Completed = 50%

- 50% of your grade depends on how long your project is compared to the other group projects. The projects that are longest will get more points. Top scores get +50%, and those following get +40%, +30%, or +20%.

- *NOTE: There is a -5% penalty for every 1/2 inch your project is out of the specifications.*

## NOTES:

**CATEGORIES:** Cantilevers, Length, Materials Strength, Paper

## MISSION: Cantilever Catastrophe II - Plastic Straws

**BRIEF:** You and your team have been selected to make the longest cantilever possible from just tape and plastic straws.

### MISSION RULES:

1. Your cantilever must between 2 and 6 inches wide, and 2 to 12 inches tall, but as long as possible. If you are outside these height and width measurements by more than 1/2 inch, you will be penalized.

2. You will work with one or two partners. Teams may not be of more than 3 people.

3. You must only use plastic straws and tape for your project.

4. The cantilever may be taped to the desk or table on one end.

5. A beam of sorts will project away from the table horizontally as far as possible. The beam distance will only be counted as far as it goes out from the table edge, provided it does not drop more than 2 inches below the table's surface. The distance will only count up to the point in the beam before it descends the acceptable level.

*TEACHER'S NOTES: For a more difficult task, add a small weight to the end of the beam, like a penny or large paper clip.*

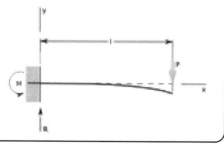

### QUIZ GRADE:

A research paper on cantilevers.

- 2-4 pictures of cantilevers 25%

- A labeled sketch and detailed concept idea based on your cantilevers pictures 50%

- Conclusions and reflections based on your results 25%

### TEST GRADE:

Your completed design and the results of the test.

- Project Completed = 50%

- 50% of your grade depends on how long your project is compared to the other group projects. The projects that are longest will get more points. Top scores get +50%, and those following get +40%, +30%, or +20%.

- *NOTE: There is a -5% penalty for every 1/2 inch your project is out of the specifications.*

### NOTES:

**CATEGORIES:** Cantilevers, Length, Materials Strength, Plastic Straws

# MISSION: Cantilever Catastrophe III - Pipe Cleaners

## BRIEF:

You and your team have been selected to make the longest cantilever possible from just tape and plastic straws.

## MISSION RULES:

1. Your cantilever must between 2 and 6 inches wide, and 2 to 12 inches tall, but as long as possible. If you are outside these height and width measurements by more than 1/2 inch, you will be penalized.

2. You will work with one or two partners. Teams may not be of more than 3 people.

3. You must only use pipe cleaners and tape for your project. Your teacher will determine how many materials you can use.

4. The cantilever may be taped to the desk or table on one end.

5. A beam of sorts will project away from the table horizontally as far as possible. The beam distance will only be counted as far as it goes out from the table edge, provided it does not drop more than 2 inches below the table's surface. The distance will only count up to the point in the beam before it descends the acceptable level.

*TEACHER'S NOTES: For a more difficult task, add a small weight to the end of the beam, like a penny or large paper clip.*

## QUIZ GRADE:

A research paper on cantilevers.

- 2-4 pictures of cantilevers 25%

- A labeled sketch and detailed concept idea based on your cantilevers pictures 50%

- Conclusions and reflections based on your results 25%

## TEST GRADE:

Your completed design and the results of the test.

- Project Completed = 50%

- 50% of your grade depends on how long your project is compared to the other group projects. The projects that are longest will get more points. Top scores get +50%, and those following get +40%, +30%, or +20%.

- *NOTE: There is a -5% penalty for every 1/2 inch your project is out of the specifications.*

## NOTES:

## CATEGORIES: Cantilevers, Length, Materials Strength, Pipe Cleaners

## MISSION: Cantilever Catastrophe IV - Wood

**BRIEF:** You and your team have been selected to make the longest cantilever possible from just toothpicks or popsicle sticks, tape, and glue.

### MISSION RULES:

1. Your cantilever must between 2 and 6 inches wide, and 2 to 12 inches tall, but as long as possible. If you are outside these height and width measurements by more than 1/2 inch, you will be penalized.

2. You will work with one or two partners. Teams may not be of more than 3 people.

3. You must only use toothpicks or popsicle sticks for your project along with tape or glue. Your teacher will determine how many materials you can use.

4. The cantilever may be taped to the desk or table on one end.

5. A beam of sorts will project away from the table horizontally as far as possible. The beam distance will only be counted as far as it goes out from the table edge, provided it does not drop more than 2 inches below the table's surface. The distance will only count up to the point in the beam before it descends the acceptable level.

*TEACHER'S NOTES: For a more difficult task, add a small weight to the end of the beam, like a penny or large paper clip.*

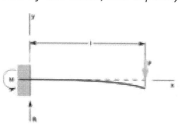

### QUIZ GRADE:

A research paper on cantilevers.

- 2-4 pictures of cantilevers 25%

- A labeled sketch and detailed concept idea based on your cantilevers pictures 50%

- Conclusions and reflections based on your results 25%

### TEST GRADE:

Your completed design and the results of the test.

- Project Completed = 50%

- 50% of your grade depends on how long your project is compared to the other group projects. The projects that are longest will get more points. Top scores get +50%, and those following get +40%, +30%, or +20%.

- *NOTE: There is a -5% penalty for every 1/2 inch your project is out of the specifications.*

### NOTES:

**CATEGORIES:** Cantilevers, Length, Materials Strength, Popsicle Sticks, Toothpicks

## MISSION: Cantilever Catastrophe V - Foil

**BRIEF:** You and your team have been selected to make the longest cantilever possible from just foil and tape.

### MISSION RULES:

1. Your cantilever must between 2 and 6 inches wide, and 2 to 12 inches tall, but as long as possible. If you are outside these height and width measurements by more than 1/2 inch, you will be penalized.

2. You will work with one or two partners. Teams may not be of more than 3 people.

3. You must only use foil for your project. Your teacher will determine how many materials you can use.

4. The cantilever may be taped to the desk or table on one end.

5. A beam of sorts will project away from the table horizontally as far as possible. The beam distance will only be counted as far as it goes out from the table edge, provided it does not drop more than 2 inches below the table's surface. The distance will only count up to the point in the beam before it descends the acceptable level.

*TEACHER'S NOTES: For a more difficult task, add a small weight to the end of the beam, like a penny or large paper clip.*

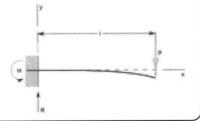

### QUIZ GRADE:

A research paper on cantilevers.

- 2-4 pictures of cantilevers 25%

- A labeled sketch and detailed concept idea based on your cantilevers pictures 50%

- Conclusions and reflections based on your results 25%

### TEST GRADE:

Your completed design and the results of the test.

- Project Completed = 50%

- 50% of your grade depends on how long your project is compared to the other group projects. The projects that are longest will get more points. Top scores get +50%, and those following get +40%, +30%, or +20%.

- *NOTE: There is a -5% penalty for every 1/2 inch your project is out of the specifications.*

### NOTES:

### CATEGORIES: Cantilevers, Foil, Length, Materials Strength

## MISSION: Claymore Tower

**BRIEF:** You and your team have been selected to make a the tallest tower possible from clay.

### MISSION RULES:

1. You will design a tower.

2. Your device may be any dimensions, but it must be as tall as possible. It must have its own base.

3. You will work with one or two partners. Teams may not be of more than 3 people.

4. You may use any shaping tools you need, but you will only get a specific amount of clay.

5. The tower may not be braced against or attached to any other objects, except for whatever table surface it is built on. Otherwise, it must be completely freestanding.

### QUIZ GRADE:

A blueprint design of your idea

- Sketch 25%

- Sketch is labeled 25%

- Explanation of strategy 25%

- Conclusions and reflections based on your results 25%

### TEST GRADE:

Your completed design and the results of the test.

- Project Completed = 50%

- 50% of your grade depends on how tall your project is compared to the other group's projects. The projects that do best will get more points.

### NOTES:

**CATEGORIES:** Clay, Height, Towers

## MISSION: Clean Sweep

**BRIEF:** You and your team have been selected to make the strongest bridge possible.

### MISSION RULES:

1. You will research bridges and get ideas for a concept for your bridge design.

2. Your bridge must be 18 inches long, between 2 and 6 inches wide, and 2 to 12 inches tall. If you are outside these measurements by more than 1/2 inch, you will be penalized.

3. You will work with two or three partners. Teams may not be of more than 4 people.

4. You must only use pipe cleaners for your project. You teacher will determine how many you get.

5. The bridge must have a place in the center of it where a tray can be attached to hold weight. Projects that hold more weight score better.

### QUIZ GRADE:

A research paper on bridges.

- 2-4 pictures of bridges 25%

- A concept idea based on your bridge pictures 50%

- Conclusions and reflections based on your results 25%

### TEST GRADE:

Your completed design and the results of the test.

- Project Completed = 50%

- 50% of your grade depends on how much weight your project holds compared to the other group's projects. The projects that do best will get more points.

- *NOTE: There is a -5% penalty for every 1/2 inch your project is out of the specifications.*

### NOTES:

**CATEGORIES:** Bridges, Materials Strength, Pipe Cleaners, Weight

## MISSION: Coin Collection

**BRIEF:** You and your team have been selected to make a device that sorts 3 types of coins into different containers.

### MISSION RULES:

1. You will design a sorting device.

2. Your device may be any shape or size, but it must sort the 3 types of coins into 3 different containers, and they must be properly sorted.

3. You will work with one or two partners. Teams may not be of more than 3 people.

4. You may use any approved materials you can find at school or at home, including paper clips, pipe cleaners, plastic straws, notecards, tape...

*TEACHER'S NOTES: Use 10 coins of assorted varieties. Quarters, Nickels, and Pennies are suggested for the variety of sizes and thicknesses. If this proves to be too hard, switch it to 2 kinds of coins, probably pennies and quarters.*

### QUIZ GRADE:

Create a blueprint design for your ideas

- Sketch 25%

- Sketch is labeled 25%

- Explanation of strategies 25%

- Conclusions and reflections based on your results 25%

### TEST GRADE:

Your completed design and the results of the test.

- Project Completed = 50%

- 50% of your grade depends on the success of your sorting. Each coin is worth 5%, and you must sort 10 coins.

### NOTES:

**CATEGORIES:** Coins, Scavengers, Sorting

## MISSION: Construction Chaos II - Crane

**BRIEF:** You and your team have been selected to make an crane to lift and move objects.

### MISSION RULES:

1. You will design a stationary crane to pick up a load and move it to a designated location.

2. Your machine may be any dimensions, but it must must be able to pick up the loads specified by your teacher.

3. You may not use your hands to directly touch the payload. You may use your hands to manipulate levers, pulleys, and other parts of your machine to complete your tasks, though.

4. There will be 3 loads. Each of them may vary in dimensions and weights. You must design a versatile machine that can scoop/ hook/pick up the loads.

5. You will work with one or two partners. Teams may not be of more than 3 people.

6. You may use any approved materials you can find at school or at home, including paper clips, pipe cleaners, plastic straws, notecards, tape...

*TEACHER'S NOTES: All 3 objects should have a hook or loop on them that can be caught. Varying weights will test the strength and durability of the machines.*

### QUIZ GRADE:

Research and design on construction equipment designs.

- A paragraph on the parts and designs of construction equipment 25%

- A concept idea for your machine sketched and explained 50%

- Conclusions and reflections based on your results 25%

### TEST GRADE:

Your completed design and the results of the test.

- Project Completed = 25%

- 75% of your grade depends on if your project actually works. Each successful delivery of the load is worth 25%

### NOTES:

### CATEGORIES: Accuracy, Gears, Levers, Pulleys, Scavengers

## MISSION: Construction Chaos I - Excavator

**BRIEF:** You and your team have been selected to make an excavator machine to transport objects.

### MISSION RULES:

1. You will design an excavator truck (backhoe, bucket truck...) to pick up a load and move it to a designated location.

2. Your machine may be any dimensions, but it must roll and it must be able to pick up the loads specified by your teacher.

3. You may not use your hands to directly touch the payload. You may use your hands to roll your machine and manipulate levers, pulleys, and other parts of your machine to complete your tasks, though.

4. There will be 3 loads. Each of them may vary in dimensions and weights. You must design a versatile machine that can scoop/ pick up the loads.

5. You will work with one or two partners. Teams may not be of more than 3 people.

6. You may use any approved materials you can find at school or at home, including paper clips, pipe cleaners, plastic straws, notecards, tape...

*TEACHER'S NOTES: Generally, moving a small load, like a AA battery, a coin, or a paperclip from one end of the table to the other without being touched by hands is the goal. Test 3 different loads or the same load 3 times.*

### QUIZ GRADE:

Research and design on construction equipment designs.

- A paragraph on the parts and designs of construction equipment 25%

- A concept idea for your machine sketched and explained 50%

- Conclusions and reflections based on your results 25%

### TEST GRADE:

Your completed design and the results of the test.

- Project Completed = 25%

- 75% of your grade depends on if your project actually works. Each successful delivery of the load is worth 25%

### NOTES:

### CATEGORIES: Accuracy, Gears, Levers, Pulleys, Scavengers

## MISSION: Construction Chaos III - Forklift

**BRIEF:** You and your team have been selected to make an forklift machine to transport objects.

### MISSION RULES:

1. You will design forklift to pick up a load and move it to a designated location.

2. Your machine may be any dimensions, but it must roll and it must be able to pick up the loads specified by your teacher.

3. You may not use your hands to directly touch the payload. You may use your hands to roll your machine and manipulate levers, pulleys, and other parts of your machine to complete your tasks, though.

4. There will be 3 loads. Each of them may vary in dimensions and weights. You must design a versatile machine that can scoop/ pick up the loads.

5. Each load MUST be picked up, not just skidded and pushed across the table surface.

6. You will work with one or two partners. Teams may not be of more than 3 people.

7. You may use any approved materials you can find at school or at home, including paper clips, pipe cleaners, plastic straws, notecards, tape...

*TEACHER'S NOTES: Small, flat objects work best for loads. Little memo pads, a baseball card, etc...*

### QUIZ GRADE:

Research and design on construction equipment designs.

- A paragraph on the parts and designs of construction equipment 25%

- A concept idea for your machine sketched and explained 50%

- Conclusions and reflections based on your results 25%

### TEST GRADE:

Your completed design and the results of the test.

- Project Completed = 25%

- 75% of your grade depends on if your project actually works. Each successful delivery of the load is worth 25%

### NOTES:

### CATEGORIES: Accuracy, Gears, Levers, Pulleys, Scavengers

# MISSION: Copper Road

**BRIEF:** You and your team have been selected to make the longest, trickiest course in which to deliver a penny to its final destination. The longer and trickier the better.

## MISSION RULES:

1.  You will design a roller coaster for a penny out of tape, paper, card stock, other supplied materials, and materials you can find at home that are approved for use.

2.  The roller coaster must successfully deliver 3 pennies, one at a time, to a collection point at the base of the project. Each penny that fails to make it to the end of the course will result in a penalty for the total score.

3.  You may test at home. In fact, you're encouraged to test at home! Assembly and design may also take place at school, but time is limited.

4.  Teams may be of no more than 4 people.

5.  Suggested Tricks to include are: stairs, vertical loops, jumps, horizontal spirals, switchbacks, tunnels, funnels, trap doors, and drain pans. Be original designing tricks. Tricks can be anything where the penny just isn't rolling in a straight path.

*TEACHER'S NOTES: Some hints should be given about the way a penny rolls and slides that can be translated into designs.*

## QUIZ GRADE:

A decorated advertisement poster that explains your concept, names your roller coaster, lists the features of the coaster, and explains which team member was responsible for each part.

- Coaster Name 10%

- Coaster Concept 25%

- List of Features 25%

- Sketch/Artwork 30%

- Who did What? 10%

## TEST GRADE:

Your completed design and the results of the test.

- Project Completed = 25%

- 1% per second of average time (average of all 3 tests)

- -5% for each penny that fails to make it to the end of the track. You will get 1 restart or 'nudge' per penny for free, but then it costs you.

- +10% per different trick, or +5% for a second trick of the same kind already used. No points for more one of the same kind of trick.

**CATEGORIES:** Coins, Paper, Time, Tracks

# MISSION: Discus

**BRIEF:** You and your team have been selected to make a device that can throw a coin as far as possible.

## MISSION RULES:

1. You will design a device that launches a coin as far as possible.

2. Your building materials should consist of paper, card stock, glue, rubber bands, popsicle sticks, and other readily-available materials that your teacher approves.

3. Your device may be no longer than 24 inches in any dimension, and no more than 48 inches total dimensions (l + w + h).

4. You will work with one or two partners. Teams may be of no more than 3 people.

5. Up to three tests will be made. Your teacher will determine how you are scored: longest shot, average, or total.

*TEACHER'S NOTES: Some care should be taken during test, so people are not injured by flying coins.*

## QUIZ GRADE:

A blueprint design of your idea

- Sketch 25%

- Sketch is labeled 25%

- Explanation of strategy 25%

- Conclusions and reflections based on your results 25%

## TEST GRADE:

Your completed design and the results of the test.

- Project Completed = 50%

- 50% of your grade depends on how far your coins travels.

- *NOTE: The best project gets an automatic 100%.*

## NOTES:

**CATEGORIES:** Coins, Distance, Throwers

# MISSION: Frisbees

**BRIEF:** You and your team have been selected to make a flying disk that can fly as far as possible.

## MISSION RULES:

1.  You will design a flying disk or ring from common classroom materials.

2.  Your building materials should consist of paper, card stock, glue, rubber bands, popsicle sticks, plastic straws, and other readily-available materials that your teacher approves.

3.  Your device should have a diameter of less than 18 inches.

4.  You will work with alone or with a single partner. Teams may be of no more than 2 people.

5.  Up to three tests will be made. Your teacher will determine how you are scored: longest shot, average, or total.

## QUIZ GRADE:

A blueprint design of your idea

- Sketch 25%

- Sketch is labeled 25%

- Explanation of strategy 25%

- Conclusions and reflections based on your results 25%

## TEST GRADE:

Your completed design and the results of the test.

- Project Completed = 50%

- 50% of your grade depends on how far your disk travels.

- *NOTE: The best project gets an automatic 100%.*

## NOTES:

**CATEGORIES:** Distance, Fliers

# MISSION: Glass Bottom Boats

**BRIEF:** You and your team have been selected to make a boat from pipe cleaners and plastic wrap to help your marshmallow family cross a lake!

## MISSION RULES:

1. You will design a boat that built entirely of only plastic wrap and pipe cleaners.

2. Your boat must have an area to house a family of four mini marshmallow family members (built from 2-3 mini marshmallows and toothpicks).

3. You will work with a single partner. Teams may not be of more than 2 people.

4. The boat must have at least one dimension greater than 6 inches.

5. The boat will be pushed across the 'lake' or tub of water with a fan, so you may want to develop some sort of sail or wind catcher.

6. Your boat must not tip over or get the family wet!

*TEACHER'S NOTES: You will need some sort of body of water. A large sink, a water tub, or even a giant trash can filled with water could work. A rain set of 1-3 rain gutters coupled together with caps at the end may hold water, too.*

*You also need a small fan to blow the boats across the water.*

## QUIZ GRADE:

Create a blueprint designs for your ideas

- Sketch 25%

- Sketch is labeled 25%

- Explanation of strategies 25%

- Conclusions and reflections based on your results 25%

## TEST GRADE:

Your completed design and the results of the test.

- Project Completed = 40%

60% of your grade depends on if your family members stay dry. Each one is worth 15%.

*NOTE: You may be required to make multiple runs across the lake, too!*

## NOTES:

**CATEGORIES:** Boats, Marshmallows, Pipe Cleaners, Plastic Wrap, Water, Wind

## MISSION: Golden Arches I - Paper

**BRIEF:** You and your team have been selected to design an arch that is as tall and wide as possible.

### MISSION RULES:

1. You will design an arch that is as tall and as wide as possible.

2. Your finished project must be built of only paper and tape.

3. Your project must be completely freestanding and may not be attached to the floor or a table surface.

4. Your teacher will determine your materials limit for the arch.

5. You will work with one or two partners. Teams may be of no more than 3 people.

6. While a perfect curve is hard to attain, some semblance of an arch must be created. A simple V or single-pitched roof is not acceptable.

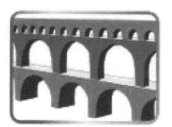

### QUIZ GRADE:

Create a blueprint design for your ideas

• Sketch 25%

• Sketch is labeled 25%

• Explanation of strategies 25%

• Conclusions and reflections based on your results 25%

### TEST GRADE:

Your completed design and the results of the test.

• Project Completed = 50%

• 50% of your project's score depends on the height and width of your project as compared to other projects.

• A formula of width X height will be used. Width is at the widest part near the base, and height is as the peak of the arch.

• *NOTE: The arch with the highest score will automatically get 100%*

### NOTES:

**CATEGORIES:** Arches, Height, Paper

# MISSION: Golden Arches II - Plastic Straws

## BRIEF:
You and your team have been selected to design an arch that is as tall and wide as possible.

## MISSION RULES:

1. You will design an arch that is as tall and as wide as possible.

2. Your finished project must be built of only plastic straws, tape, and a small amount of card stock.

3. Your project must be completely freestanding and may not be attached to the floor or a table surface.

4. Your teacher will determine your materials limit for the arch.

5. You will work with one or two partners. Teams may be of no more than 3 people.

6. While a perfect curve is hard to attain, some semblance of an arch must be created. A simple V or single-pitched roof is not acceptable.

## QUIZ GRADE:

Create a blueprint design for your ideas

- Sketch 25%

- Sketch is labeled 25%

- Explanation of strategies 25%

- Conclusions and reflections based on your results 25%

## TEST GRADE:

Your completed design and the results of the test.

- Project Completed = 50%

- 50% of your project's score depends on the height and width of your project as compared to other projects.

- A formula of width X height will be used. Width is at the widest part near the base, and height is as the peak of the arch.

- *NOTE: The arch with the highest score will automatically get 100%*

## NOTES:

## CATEGORIES: Arches, Height, Plastic Straws

# MISSION: Golden Arches III - Foil

**BRIEF:** You and your team have been selected to design an arch that is as tall and wide as possible.

## MISSION RULES:

1. You will design an arch that is as tall and as wide as possible.

2. Your finished project must be built of only foil and a small amount of tape.

3. Your project must be completely freestanding and may not be attached to the floor or a table surface.

4. Your teacher will determine your materials limit for the arch.

5. You will work with one or two partners. Teams may be of no more than 3 people.

6. While a perfect curve is hard to attain, some semblance of an arch must be created. A simple V or single-pitched roof is not acceptable.

## QUIZ GRADE:

Create a blueprint design for your ideas

- Sketch 25%

- Sketch is labeled 25%

- Explanation of strategies 25%

- Conclusions and reflections based on your results 25%

## TEST GRADE:

Your completed design and the results of the test.

- Project Completed = 50%

- 50% of your project's score depends on the height and width of your project as compared to other projects.

- A formula of width X height will be used. Width is at the widest part near the base, and height is as the peak of the arch.

- *NOTE: The arch with the highest score will automatically get 100%*

## NOTES:

**CATEGORIES:** Arches, Foil, Height

# MISSION: Golden Arches IV - Pipe Cleaners

**BRIEF:** You and your team have been selected to design an arch that is as tall and wide as possible.

## MISSION RULES:

1. You will design an arch that is as tall and as wide as possible.

2. Your finished project must be built of only pipe cleaners.

3. Your project must be completely freestanding and may not be attached to the floor or a table surface.

4. Your teacher will determine your materials limit for the arch.

5. You will work with one or two partners. Teams may be of no more than 3 people.

6. While a perfect curve is hard to attain, some semblance of an arch must be created. A simple V or single-pitched roof is not acceptable.

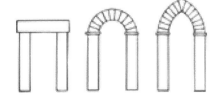

## QUIZ GRADE:

Create a blueprint design for your ideas

- Sketch 25%

- Sketch is labeled 25%

- Explanation of strategies 25%

- Conclusions and reflections based on your results 25%

## TEST GRADE:

Your completed design and the results of the test.

- Project Completed = 50%

- 50% of your project's score depends on the height and width of your project as compared to other projects.

- A formula of width X height will be used. Width is at the widest part near the base, and height is as the peak of the arch.

- *NOTE: The arch with the highest score will automatically get 100%*

## NOTES:

**CATEGORIES:** Arches, Height, Pipe Cleaners

## MISSION: Golden Arches V - Wood

**BRIEF:** You and your team have been selected to design an arch that is as tall and wide as possible.

### MISSION RULES:

1. You will design an arch that is as tall and as wide as possible.

2. Your finished project must be built of only toothpicks and/or popsicle sticks and glue.

3. Your project must be completely freestanding and may not be attached to the floor or a table surface.

4. Your teacher will determine your materials limit for the arch.

5. You will work with one or two partners. Teams may be of no more than 3 people.

6. While a perfect curve is hard to attain, some semblance of an arch must be created. A simple V or single-pitched roof is not acceptable.

### QUIZ GRADE:

Create a blueprint design for your ideas

- Sketch 25%

- Sketch is labeled 25%

- Explanation of strategies 25%

- Conclusions and reflections based on your results 25%

### TEST GRADE:

Your completed design and the results of the test.

- Project Completed = 50%

- 50% of your project's score depends on the height and width of your project as compared to other projects.

- A formula of width X height will be used. Width is at the widest part near the base, and height is as the peak of the arch.

- *NOTE: The arch with the highest score will automatically get 100%*

### NOTES:

**CATEGORIES:** Arches, Height, Popsicle Sticks, Toothpicks

## MISSION: Golden Arches VI - Clay

**BRIEF:** You and your team have been selected to design an arch that is as tall and wide as possible.

### MISSION RULES:

1. You will design an arch that is as tall and as wide as possible.

2. Your finished project must be built of only clay.

3. Your project must be completely freestanding and may not be attached to the floor or a table surface.

4. Your teacher will determine your materials limit for the arch.

5. You will work with one or two partners. Teams may be of no more than 3 people.

6. While a perfect curve is hard to attain, some semblance of an arch must be created. A simple V or single-pitched roof is not acceptable.

### QUIZ GRADE:

Create a blueprint design for your ideas

- Sketch 25%

- Sketch is labeled 25%

- Explanation of strategies 25%

- Conclusions and reflections based on your results 25%

### TEST GRADE:

Your completed design and the results of the test.

- Project Completed = 50%

- 50% of your project's score depends on the height and width of your project as compared to other projects.

- A formula of width X height will be used. Width is at the widest part near the base, and height is as the peak of the arch.

- *NOTE: The arch with the highest score will automatically get 100%*

### NOTES:

**CATEGORIES:** Arches, Clay, Height

## MISSION: The Irrigator

### BRIEF:

You and your team have been selected to make the longest, trickiest course in which to deliver water to its final destination. The longer and trickier the better.

### MISSION RULES:

1. You will design a roller coaster for water made out of tape, plastic, clay, tubing, straws, styrofoam, wood, and materials you can find at home that are approved for use.

2. The roller coaster must successfully deliver 1 cup of water to a collection point at the base of the project. Water that leaks or fails to make it to the collection point will cost you points.

3. You may test at home. In fact, you're encouraged to test at home! Assembly and design may also take place at school, but time is limited.

4. Teams may be of no more than 4 people.

5. Suggested Tricks to include are: stairs, vertical loops, jumps, horizontal spirals, switchbacks, tunnels, funnels, trap doors, and drain pans. Be original designing tricks. Tricks can be anything where the water just isn't rolling in a straight path.

*TEACHER'S NOTES: Adding color to the water makes it easier to track leaks and adds excitement. Maybe use red for trial 1, blue for trial 2, and green for trial 3.*

### QUIZ GRADE:

A decorated advertisement poster that explains your concept, names your roller coaster, lists the features of the coaster, and explains which team member was responsible for each part.

- Coaster Name 10%

- Coaster Concept 25%

- List of Features 25%

- Sketch/Artwork 30%

- Who did What? 10%

### TEST GRADE:

Your completed design and the results of the test.

- Project Completed = 25%

- 5% per second of average time (average of all 3 tests)

- -5% for each ounce of water that fails to make it through the track.

- +10% per different trick, or +5% for a second trick of the same kind already used. No points for more one of the same kind of trick.

### CATEGORIES: Plastic Straws, Scavengers, Time, Tracks, Water

# MISSION: Kitetastrophe!

**BRIEF:** You and your team have been selected to make a variety of kites using different materials.

## MISSION RULES:

1. You will design 3 kites.

2. You will work with one or two partners. Teams may not be of more than 3 people.

3. Kites should be designed using different materials like: foil, wax paper, plastic wrap, tissue paper, or cooking parchment. The frames can be assembled with bamboo skewers, plastic straws, or other lightweight materials readily available.

4. All 3 kites must take advantage of different building materials.

5. There are no size or shape restrictions, but they MUST fly!

## QUIZ GRADE:

Create blueprint designs for your ideas

- Sketches 25%

- Sketches are labeled 25%

- Explanation of strategies 25%

- Conclusions and reflections based on your results 25%

## TEST GRADE:

Your grade relies completely on your test results!

- Each kite is worth 33% of your grade. Each one that flies gets you that many points! Each one that does not fly costs you those 33%.

- If you get all 3 airborne for at least 10 seconds (without running to keep it aloft), you get the bonus 1% to make 100% :)

## NOTES:

**CATEGORIES:** Fliers, Foil, Height, Paper, Plastic Wrap, String, Wax Paper, Wind

# MISSION: Lead Sinker

**BRIEF:** You and your team have been selected to design a device that can sink a ping pong ball and to make it as small as possible.

## MISSION RULES:

1. You will design a device from scavenged materials. Materials must be approved by the teacher.

2. The device must sink the ping pong ball and stay sunk for at least 10 seconds.

3. You will work with one or two partners. Teams may be of no more than 3 people.

4. You may get test trials to see if your device works. The number of test runs depends on your teacher.

5. Your device must have a cup or receiver of some sort to hold the ping pong ball. The ping pong ball must be at least partially visible at all times.

6. Your device may not have any dimension larger than 6 inches.

*TEACHER'S NOTES: You may want to disallow very heavy materials. If this seems too easy, decrease the dimensions allowed or increase the number of ping pong balls.*

## QUIZ GRADE:

A blueprint design of your idea

- Sketch 25%

- Sketch is labeled 25%

- Explanation of strategy 25%

- Conclusions and reflections based on your results 25%

## TEST GRADE:

Your completed design and the results of the test.

- Project Completed = 50% for successfully sinking a ping pong ball.

- 50% of your grade depends on how small your project is compared to the others. You will be awarded these points in 10% increments.

- *OPTIONAL: Points may be awarded differently for tests requiring you to sink more than one ping pong ball.*

## NOTES:

**CATEGORIES:** Boats, Buoyancy, Ping Pong Balls, Scavengers, Water, Weight

## MISSION: Leaning Tower of Pasta

**BRIEF:** You and your team have been selected to make a the tallest tower possible from spaghetti noodles and marshmallows.

### MISSION RULES:

1. You will design a tower.

2. Your device may be any dimensions, but it must be as tall as possible. It must have its own base.

3. You will work with one or two partners. Teams may not be of more than 3 people.

4. You may use only spaghetti and marshmallows to build your project. Your teacher will determine your maximum amount of materials,

5. The tower may not be braced against or attached to any other objects, except for whatever table surface it is built on. Otherwise, it must be completely freestanding.

### QUIZ GRADE:

A blueprint design of your idea

• Sketch 25%

• Sketch is labeled 25%

• Explanation of strategy 25%

• Conclusions and reflections based on your results 25%

### TEST GRADE:

Your completed design and the results of the test.

• Project Completed = 50%

• 50% of your grade depends on how tall your project is compared to the other group's projects. The projects that do best will get more points.

### NOTES:

**CATEGORIES:** Height, Marshmallows, Pasta, Towers

## MISSION: Look Out Below!

**BRIEF:** You and your team have been selected to make a parachute that slows a weight down as much as possible, taking the longest time to reach the floor from the drop point.

## MISSION RULES:

1.  You will design a device to slow the descent of a weight your teacher has chosen (a graduated weight, a ping pong ball, army man, etc...) as much as possible

2.  Your device may be built from any approved materials you can find at home or in the class, including: straws, tissue paper, tape or glue, foil, strings notecards...

3.  You will work with one or two partners. Teams may be of no more than 3 people.

4.  Your device will be tested from a set height of 6-10 feet, or higher if possible.

5.  Your device must have some sort of hook or basket for the weight to be attached to your project. The requirements for this depend on what weighted object your teacher has selected.

6.  There will be up to 3 tests with small adjustments allowed between testings.

*TEACHER'S NOTES: Use of a stopwatch and a ladder may make this project better. Awards can be given for slowest time (which is best) and best average.*

## QUIZ GRADE:

Create a blueprint designs for your ideas

*   Sketch 25%

*   Sketch is labeled 25%

*   Explanation of strategies 25%

*   Conclusions and reflections based on your results 25%

## TEST GRADE:

Your completed design and the results of the test.

*   Project Completed = 50%

*   50% of your grade depends on how long your project stays in the air compared to the other group's projects. The projects that do best will get more points.

*   *NOTE: The project that takes the longest to land gets an automatic 100%*

## NOTES:

**CATEGORIES:** Fliers, Scavengers, Time

## MISSION: Lost Your Marbles

**BRIEF:** You and your team have been selected to make a device that sorts marbles and ping pong balls into different containers.

### MISSION RULES:

1. You will design a sorting device.

2. Your device may be any shape or size, but it must sort the 2 types of balls into 2 different containers, and they must be properly sorted.

3. You will work with one or two partners. Teams may not be of more than 3 people.

4. You may use any approved materials you can find at school or at home, including paper clips, pipe cleaners, plastic straws, notecards, tape...

*TEACHER'S NOTES: Use 5 total marbles and ping pong balls.*

### QUIZ GRADE:

Create a blueprint design for your ideas

• Sketch 25%

• Sketch is labeled 25%

• Explanation of strategies 25%

• Conclusions and reflections based on your results 25%

### TEST GRADE:

Your completed design and the results of the test.

• Project Completed = 50%

• 50% of your grade depends on the success of your sorting. Each ball is worth 10%, and you must sort 5 balls or marbles.

### NOTES:

**CATEGORIES:** Marbles, Ping Pong Balls, Scavengers, Sorting

## MISSION: Make it Rain!

**BRIEF:** You and your team have been selected to make a device to throw a water balloon as far as possible.

### MISSION RULES:

1. You will design a throwing device.

2. Your device must be no longer than 18 inches, no taller than 18 inches, and no wider than 12 inches when assembled and stationed at the throwing line.

3. You will work with two or three partners. Teams may not be of more than 4 people.

4. You must only use paper, glue, tape, rubber bands, paperclips, pencils, popsicle sticks, or other approved office supplies for your project.

5. The device must have some cup or place to put the small water balloon. The device will then be manipulated and the attempt measured.

*TEACHER'S NOTES: 3 attempts are suggested. A total distance category and/or an average category could be considered for special honors. Mini water balloons are highly suggested, as is doing this outside!*

### QUIZ GRADE:

Create a blueprint design for your ideas

- Sketch 25%

- Sketch is labeled 25%

- Explanation of strategies 25%

- Conclusions and reflections based on your results 25%

### TEST GRADE:

Your completed design and the results of the test.

- Project Completed = 50%

- 50% of your grade depends on how far your project throws compared to the other group's projects. The projects that do best will get more points.

- *NOTE: There is a -5% penalty for every 1/2 inch your project is out of the size specifications.*

### NOTES:

**CATEGORIES:** Balloons, Distance, Popsicle Sticks, Rubber Bands, Throwers, Water

## MISSION: Marshmallow Mayhem

**BRIEF:** You and your team have been selected to make a device that can crush marshmallows as flat as possible.

### MISSION RULES:

1. You will design a device that will crush a regular-sized marshmallow as flat as possible.

2. 3 seconds after being hit/crushed/smashed, the marshmallow will be measured for height at the tallest point.

3. You may use any approved materials that you find in the classroom or at home.

4. You may not simply drop a weight on the marshmallow. There must be a device/lever/mechanism that creates an action that crushes the marshmallow.

5. You will work with one to two partners. Teams may not be of more than 3 people.

6. 3 tests will be made. Your teacher may decide on a best of 3, an average score, or some composite score.

### QUIZ GRADE:

Create a blueprint design for your ideas

- Sketch 25%

- Sketch is labeled 25%

- Explanation of strategies 25%

- Conclusions and reflections based on your results 25%

### TEST GRADE:

Your completed design and the results of the test.

- Project Completed = 50%

- 50% of your grade depends on how flattened your marshmallow is compared to others.

- Top scores get +50%, and those following get +40%, +30%, or +20%.

### NOTES:

**CATEGORIES:** Crashes, Marshmallows, Scavengers

## MISSION: Marshmallow Snowman Pileup

**BRIEF:** You and your team have been selected to make an MPD (marshmallow protection device). This device must be a rolling vehicle that can protect a marshmallow family as it goes down an obstacle course or simple track your teacher has developed. Track conditions MAY vary.

### MISSION RULES:

1. You will design a rolling vehicle of dimensions your teacher requires that will fit on the track.

2. The vehicle must not have a completely closed top. At least 50% of the top half of the car must be open.

3. Your assembled marshmallow snowman family of 4 must be visible at all times and may not be glued or taped down to the car.

4. You may use any materials you want, provided you can scrounge them up, buy them, or find them.

5. You may test at home. In fact, you're encouraged to test at home! Assembly and design may also take place at school, but time is limited.

6. Teams may be of no more than 3 people.

*TEACHER'S NOTES: This one works well with holiday snowman-shaped marshmallows. Use toothpicks to add arms and legs and markers to draw on faces. Similarly, you can run a toothpick through 2-3 mini marshmallows to make your family.*

*Your track can be built with simple boards, lengths of taped styrofoam, or plastic rain gutters.*

### QUIZ GRADE:

Create a blueprint designs for your ideas

- Labeled Sketch 25%

- Materials List 25%

- Explanation of strategies 25%

- Conclusions and reflections based on your results 25%

### TEST GRADE:

Your completed design and the results of the test.

- Completed Design = 40% (penalties assessed if it does not follow the rules)

Results of Trials = up to 60%

- Each surviving family member is worth 5%.

- -5% penalty per family member that fall out of the car at any point in the track run.

- 3 trials x 4 family members = 12 lives to save, and 12 x 5% = a possible of 60%

### NOTES:

**CATEGORIES:** Cars, Crashes, Marshmallows, Tracks

## MISSION: Nice Cold Drink

**BRIEF:** You and your team have been selected to make a device that cleans water.

### MISSION RULES:

1. You will design a device to clean water.

2. Your device may be any size, but it must have a place to put dirty water into the device and a place for clean water to collect.

3. You will work with one or two partners. Teams may not be of more than 3 people.

4. You may use any approved materials you can find at school or at home, including paper clips, pipe cleaners, plastic straws, notecards, tape...

*TEACHER'S NOTES: It is suggested to have a variety of pollutants in the water of varying sizes. Dirt, sawdust, gravel, and even food coloring can be added. Better filters will catch more sizes and types of debris.*

### QUIZ GRADE:

Research and design of a water filtration device.

- A paragraph on water filtration 25%

- A concept idea for your filtration device sketched and explained 50%

- Conclusions and reflections based on your results 25%

### TEST GRADE:

Your completed design and the results of the test.

- Project Completed = 50%

- 50% of your grade depends on if your project actually works. Cleaner = better grade.

- *NOTE: The best project gets an automatic 100%.*

### NOTES:

**CATEGORIES:** Scavengers, Sorting, Water

# MISSION: The Old Swimming Raft

## BRIEF:

You and your team have been selected to design a ping pong ball raft that can hold the most paperclips without sinking.

## MISSION RULES:

1. You will design a raft with 4 ping pong balls and scavenged materials like: paper, toothpicks, tape, cardboard, plastic...

2. The boat must float. You will be given 3 chances to see if it floats prior to the real testing.

3. You will work with one or two partners. Teams may be of no more than 3 people.

4. After putting your floating boat into the water for the actual test, paperclips will be added until it sinks or tips over and dumps the paperclips.

5. Your boat must have a cup or receiver of some sort to hold the weights. If this fills up and there is no more room to add weight, no more weight will be added.

*TEACHER'S NOTE: Pennies, measured amounts of sand, or graduated weights can also be used instead of paperclips.*

## QUIZ GRADE:

A blueprint design of your idea

- Sketch 25%

- Sketch is labeled 25%

- Explanation of strategy 25%

- Conclusions and reflections based on your results 25%

## TEST GRADE:

Your completed design and the results of the test.

- Project Completed = 50% for successfully floating.

- 50% of your grade depends on how much weight the boat holds. The more it holds in comparison to the other teams, the better you do.

- You will be awarded these points in 10% increments. The boat that holds the most clips automatically gets 100%

## NOTES:

## CATEGORIES: Boats, Buoyancy, Ping Pong Balls, Scavengers, Water, Weight

# MISSION: Pin the Tail on the Balloon

## BRIEF:

You and your team have been selected to make a device that can pop balloons.

## MISSION RULES:

1. You will design a device that will pop balloons.

2. You may use any approved materials that you find in the classroom or at home.

3. You may not simply drop a weight on the balloon. There must be a device/lever/mechanism that creates an action that pops the balloon.

4. You will work with one to two partners. Teams may not be of more than 3 people.

5. 3 tests will be made. Your teacher may decide on a best of 3, an average score, or some composite score.

*TEACHER'S NOTES: Try balloons of varying levels of inflation. Less inflated balloons are often harder to pop. You will want to carefully monitor what potentially dangerous objects are used to pop the balloons. :)*

## QUIZ GRADE:

Create a blueprint design for your ideas

- Sketch 25%

- Sketch is labeled 25%

- Explanation of strategies 25%

- Conclusions and reflections based on your results 25%

## TEST GRADE:

Your completed design and the results of the test.

- Project Completed = 25%

- 75% of your grade depends on the results of your trials. Each successful test is worth 25%.

## NOTES:

## CATEGORIES: Balloons, Crashes, Scavengers

# MISSION: Ping Pong Madness

**BRIEF:** You and your team have been selected to make the longest, trickiest course in which to deliver a ping pong ball to its final destination. The longer and trickier the better.

## MISSION RULES:

1. You will design a roller coaster for a ping pong ball out of tape, cardboard tubes, paper, card stock, other supplied materials, and materials you can find at home that are approved for use.

2. The roller coaster must successfully deliver 3 ping pong balls, one at a time, to a collection point at the base of the project. Each ball that fails to make it to the end of the course will result in a penalty for the total score.

3. You may test at home. In fact, you're encouraged to test at home! Assembly and design may also take place at school, but time is limited.

4. Teams may be of no more than 4 people.

5. Suggested Tricks to include are: stairs, vertical loops, jumps, horizontal spirals, switchbacks, tunnels, funnels, trap doors, and drain pans. Be original designing tricks. Tricks can be anything where the ball just isn't rolling in a straight path.

## QUIZ GRADE:

A decorated advertisement poster that explains your concept, names your roller coaster, lists the features of the coaster, and explains which team member was responsible for each part.

• Coaster Name 10%

• Coaster Concept 25%

• List of Features 25%

• Sketch/Artwork 30%

• Who did What? 10%

## TEST GRADE:

Your completed design and the results of the test.

• Project Completed = 25%

• 1% per second of average time (average of all 3 ping pong balls tests)

• -5% for each ping pong ball that fails to make it to the end of the track. You will get 1 restart or 'nudge' per ball for free, but then it costs you.

• +10% per different trick, or +5% for a second trick of the same kind already used. No points for more one of the same kind of trick.

**CATEGORIES:** Cardboard Tubes, Paper, Ping Pong Balls, Time, Tracks

## MISSION: Ping Pong Mortar

**BRIEF:** You and your team have been selected to make a mortar launcher for a ping pong ball.

### MISSION RULES:

1. You will design a device that launches a ping pong ball as high into the air as possible.

2. Your primary building materials should be a cardboard tube(s) and some materials to help fling or launch the ping pong ball through the tube and into the air.

3. You will work with one or two partners. Teams may be of no more than 3 people.

4. Up to three tests will be made.

*TEACHER'S NOTES: This is often best done outside, where the height of the launch can be measured against how many bricks the ping pong ball flies above, graduated lines marked on the wall, or even meter sticks taped to the wall.*

*After 3 tests, honors can be given for the best average and/or the highest single shot.*

### QUIZ GRADE:

A blueprint design of your idea

- Sketch 25%

- Sketch is labeled 25%

- Explanation of strategy 25%

- Conclusions and reflections based on your results 25%

### TEST GRADE:

Your completed design and the results of the test.

- Project Completed = 50%

- 50% of your grade depends on how high your project travels.

- *NOTE: The best project gets an automatic 100%.*

### NOTES:

**CATEGORIES:** Cardboard Tubes, Height, Ping Pong Balls, Throwers

# MISSION: The Pipeworks

## BRIEF:

You and your team have been selected to make a the tallest tower possible from pipe cleaners.

## MISSION RULES:

1. You will design a tower.

2. Your device may be any dimensions, but it must be as tall as possible. It must have its own base.

3. You will work with one or two partners. Teams may not be of more than 3 people.

4. You may use only scissors and a specific amount of pipe cleaners to build your tower.

5. The tower may not be braced against or attached to any other objects. It must be completely freestanding.

## QUIZ GRADE:

A blueprint design of your idea

- Sketch 25%

- Sketch is labeled 25%

- Explanation of strategy 25%

- Conclusions and reflections based on your results 25%

## TEST GRADE:

Your completed design and the results of the test.

- Project Completed = 50%

- 50% of your grade depends on how tall your project is compared to the other group's projects. The projects that do best will get more points.

- *NOTE: There is a -5% penalty for each new sheet of paper you are exchanging for one your team made a mistake on and wasted.*

## NOTES:

## CATEGORIES: Height, Pipe Cleaners, Towers

## MISSION: Pneumatic Cannon

**BRIEF:** You and your team have been selected to make an air-powered cannon for a ping pong ball.

### MISSION RULES:

1. You will design a device that launches a ping pong ball as far as possible using air power.

2. Your primary building materials should be a cardboard tube(s) and a balloon that will launch the ping pong ball through the tube and into the air.

3. You will work with one or two partners. Teams may be of no more than 3 people.

4. Up to three tests will be made.

*TEACHER'S NOTES: This can be done with or without a bounce. If you don't want to measure the momentum of the bounces and that distance, shoot onto grass, carpet, or sand to slow the ball down. After 3 tests, honors can be given for the best average and/or the highest single shot.*

### QUIZ GRADE:

A blueprint design of your idea

- Sketch 25%

- Sketch is labeled 25%

- Explanation of strategy 25%

- Conclusions and reflections based on your results 25%

### TEST GRADE:

Your completed design and the results of the test.

- Project Completed = 50%

- 50% of your grade depends on how far your project shoots the ping pong ball.

- *NOTE: The best project gets an automatic 100%.*

### NOTES:

**CATEGORIES:** Balloons, Cardboard Tubes, Distance, Ping Pong Balls, Throwers

## MISSION: Pump You Up

**BRIEF:** You and your team have been selected to make the strongest possible book holder out of paper tubes and glue.

## MISSION RULES:

1. You will design a device at least 6 inches tall with any other dimensions of length and width that can hold up as many textbooks as possible without collapsing.

2. Your teacher will determine the maximum number of cardboard tubes you may use in your project.

3. You will work with one to two partners. Teams may not be of more than 3 people.

4. You may use only cardboard tubes and glue for your construction.

5. Your device must be free-standing and movable. It cannot be attached to any surface.

*TEACHER'S NOTES: You might want to have students start collecting cardboard tubes from paper towels and gift wrapping paper a good deal ahead of time before they start the project. Toilet paper rolls are likely not sanitary...*

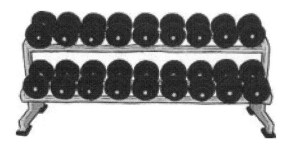

## QUIZ GRADE:

Create a blueprint design for your ideas

- Sketch 25%

- Sketch is labeled 25%

- Explanation of strategies 25%

- Conclusions and reflections based on your results 25%

## TEST GRADE:

Your completed design and the results of the test.

- Project Completed = 50%

- 50% of your grade depends on how much weight your project can hold up compared to others.

- Top scores get +50%, and those following get +40%, +30%, or +20%.

## NOTES:

**CATEGORIES:** Cardboard Tubes, Dead Lift, Materials Strength, Weight

## MISSION: Rain Marbles Down on Them!

**BRIEF:** You and your team have been selected to make a device to throw a marble as far as possible.

### MISSION RULES:

1. You will design a throwing device.

2. Your device must be no longer than 18 inches, no taller than 18 inches, and no wider than 12 inches when assembled and stationed at the throwing line.

3. You will work with two or three partners. Teams may not be of more than 4 people.

4. You must only use paper, glue, tape, rubber bands, paperclips, pencils, popsicle sticks, or other approved office supplies for your project.

5. The device must have some cup or place to put the marble. The device will then be manipulated and the attempt measured.

*TEACHER'S NOTES: 3 attempts are suggested. A total distance category and/or an average category could be considered for special honors.*

### QUIZ GRADE:

Create a blueprint design for your ideas

- Sketch 25%

- Sketch is labeled 25%

- Explanation of strategies 25%

- Conclusions and reflections based on your results 25%

### TEST GRADE:

Your completed design and the results of the test.

- Project Completed = 50%

- 50% of your grade depends on how far your project throws compared to the other group's projects. The projects that do best will get more points.

- *NOTE: There is a -5% penalty for every 1/2 inch your project is out of the size specifications.*

### NOTES:

**CATEGORIES:** Distance, Marbles, Popsicle Sticks, Rubber Bands, Throwers

# MISSION: Splashing Around

## BRIEF:

You and your team have been selected to make a water wheel that turns with moving water and performs one or more actions.

## MISSION RULES:

1. You will design and build a water wheel from notecards, glue, tape, toothpicks, popsicle sticks, rubber bands, straws, or other scavenged household materials.

2. Your finished water wheel must turn as result of moving water, like the water from a faucet.

3. Your finished water wheel must have gears or mechanisms that turn or make at least one evident and visible action occur, like raising and lowering a flag, pushing something, etc...

4. You will work with two or three partners. Teams may be of no more than 4 people.

*TEACHERS' NOTES: This one works best over by a faucet. Devices would have to be designed to sit on a counter, which a portion of the device (the water wheel and some sort of driveshaft) extending out to reach the running water.*

## QUIZ GRADE:

A blueprint design of your idea

- Sketch 25%

- Sketch is labeled 25%

- Explanation of strategy 25%

- Conclusions and reflections based on your results 25%

## TEST GRADE:

Your completed design and the results of the test.

- Project Completed = 50%

50% of your grade depends on what your project does as an action(s).

- One action = +20%

- Two actions = +35%

- Three actions = +50%

## NOTES:

## CATEGORIES: Gears, Rotation, Scavengers, Water

## MISSION: Tear the Trampoline I - Plastic Wrap

**BRIEF:** You and your team have been selected to make the strongest possible trampoline with plastic wrap!

### MISSION RULES:

1. You will design a device that holds a 12x12 inch piece of plastic wrap.

2. Your device will be set up over a gap between tables.

3. Your device will be some sort of frame designed to hold the plastic wrap sheet without tearing it as weight is added.

4. If the plastic wrap tears or is ripped out of your device, no more weight will be added, and the test will be over.

5. You will work with one to two partners. Teams may not be of more than 3 people.

6. You may use any approved materials for your product, including: popsicle sticks, glue, toothpicks, paper, tape, card stock, etc...

7. Your device must be free-standing and movable. It cannot be attached to any surface.

*TEACHER'S NOTES: Suggested weights are: marbles, pennies, or graduated weights. It might be a good idea to have a bucket to get the weights when they fall through the device.*

### QUIZ GRADE:

Create a blueprint design for your ideas

- Sketch 25%

- Sketch is labeled 25%

- Explanation of strategies 25%

- Conclusions and reflections based on your results 25%

### TEST GRADE:

Your completed design and the results of the test.

- Project Completed = 50%

- 50% of your grade depends on how much weight your project can hold up compared to others.

- Top scores get +50%, and those following get +40%, +30%, or +20%.

### NOTES:

**CATEGORIES:** Materials Strength, Plastic Wrap, Trampolines, Weight

## MISSION: Tear the Trampoline II - Wax Paper

**BRIEF:** You and your team have been selected to make the strongest possible trampoline with wax paper!

### MISSION RULES:

1. You will design a device that holds a 12x12 inch piece of wax paper.

2. Your device will be set up over a gap between tables.

3. Your device will be some sort of frame designed to hold the wax paper sheet without tearing it as weight is added.

4. If the wax paper tears or is ripped out of your device, no more weight will be added, and the test will be over.

5. You will work with one to two partners. Teams may not be of more than 3 people.

6. You may use any approved materials for your product, including: popsicle sticks, glue, toothpicks, paper, tape, card stock, etc...

7. Your device must be free-standing and movable. It cannot be attached to any surface.

*TEACHER'S NOTES: Suggested weights are: marbles, pennies, or graduated weights. It might be a good idea to have a bucket to get the weights when they fall through the device.*

### QUIZ GRADE:

Create a blueprint design for your ideas

- Sketch 25%

- Sketch is labeled 25%

- Explanation of strategies 25%

- Conclusions and reflections based on your results 25%

### TEST GRADE:

Your completed design and the results of the test.

- Project Completed = 50%

- 50% of your grade depends on how much weight your project can hold up compared to others.

- Top scores get +50%, and those following get +40%, +30%, or +20%.

### NOTES:

**CATEGORIES:** Materials Strength, Trampolines, Wax Paper, Weight

## MISSION: Tear the Trampoline III - Tissue

**BRIEF:** You and your team have been selected to make the strongest possible trampoline with tissue!

### MISSION RULES:

1. You will design a device that holds a piece of tissue paper, kleenex, or paper towel, as determined by your teacher.

2. Your device will be set up over a gap between tables.

3. Your device will be some sort of frame designed to hold the tissue without tearing it as weight is added.

4. If the tissue tears or is ripped out of your device, no more weight will be added, and the test will be over.

5. You will work with one to two partners. Teams may not be of more than 3 people.

6. You may use any approved materials for your product, including: popsicle sticks, glue, toothpicks, paper, tape, card stock, etc...

7. Your device must be free-standing and movable. It cannot be attached to any surface.

*TEACHER'S NOTES: Suggested weights are: marbles, pennies, or graduated weights. It might be a good idea to have a bucket to get the weights when they fall through the device.*

### QUIZ GRADE:

Create a blueprint design for your ideas

- Sketch 25%

- Sketch is labeled 25%

- Explanation of strategies 25%

- Conclusions and reflections based on your results 25%

### TEST GRADE:

Your completed design and the results of the test.

- Project Completed = 50%

- 50% of your grade depends on how much weight your project can hold up compared to others.

- Top scores get +50%, and those following get +40%, +30%, or +20%.

### NOTES:

**CATEGORIES:** Materials Strength, Tissue, Trampolines, Weight

# MISSION: Tear the Trampoline IV - Paper

**BRIEF:** You and your team have been selected to make the strongest possible trampoline with paper!

## MISSION RULES:

1. You will design a device that holds a piece of paper.

2. Your device will be set up over a gap between tables.

3. Your device will be some sort of frame designed to hold the paper without tearing it as weight is added.

4. If the paper tears or is ripped out of your device, no more weight will be added, and the test will be over.

5. You will work with one to two partners. Teams may not be of more than 3 people.

6. You may use any approved materials for your product, including: popsicle sticks, glue, toothpicks, paper, tape, card stock, etc...

7. Your device must be free-standing and movable. It cannot be attached to any surface.

*TEACHER'S NOTES: Suggested weights are: marbles, pennies, or graduated weights. It might be a good idea to have a bucket to get the weights when they fall through the device.*

## QUIZ GRADE:

Create a blueprint design for your ideas

- Sketch 25%

- Sketch is labeled 25%

- Explanation of strategies 25%

- Conclusions and reflections based on your results 25%

## TEST GRADE:

Your completed design and the results of the test.

- Project Completed = 50%

- 50% of your grade depends on how much weight your project can hold up compared to others.

- Top scores get +50%, and those following get +40%, +30%, or +20%.

## NOTES:

**CATEGORIES:** Materials Strength, Paper, Trampolines, Weight

## MISSION: Tubular Balls

**BRIEF:** You and your team have been selected to design a delivery device that can take a ping pong ball across the room and drop it into an open container.

### MISSION RULES:

1. You will design a delivery system to transport a ping pong ball as far as possible across the room into an open container.

2. Your finished project must be built of only paper, card stock, cardboard tubes, tape, glue, and other materials approved by your teacher.

3. You will work with one or two partners. Teams may be of no more than 3 people.

4. Your project may NOT be attached to the open container in any way.

5. Your project may use furniture as a pivot or fulcrum, but it must not be taped to the floor or furniture.

6. Your device may not be an enclosed tube for more than 50% of its length.

### QUIZ GRADE:

Create a blueprint design for your ideas

- Sketch 25%

- Sketch is labeled 25%

- Explanation of strategies 25%

- Conclusions and reflections based on your results 25%

### TEST GRADE:

Your completed design and the results of the test.

- Project Completed = 25%

- 50% of your grade depends on your success in your 3 trials. Success 3/3 = +50%, Success 2/3 = +35%, and Success 1/3 = +20%, otherwise no points.

- The final 25% of your grade is the comparative length of your project compared to the others. Longer projects that are successful get more points.

- *NOTE: The longest project that deposits all 3 balls into the container gets an automatic 100%*

### NOTES:

**CATEGORIES:** Accuracy, Cardboard Tubes, Distance, Ping Pong Balls, Tracks

**(C) 2014 Andrew Frinkle**

## MISSION: Wheel of Fortune

**BRIEF:** You and your team have been selected to make a rotating ferris wheel that is powered by rubber bands.

### MISSION RULES:

1. You will design a ferris wheel that is powered by rubber bands.

2. Your ferris wheel must be at least 6 inches in diameter.

3. You will work with one or two partners. Teams may not be of more than 3 people.

4. You may use any approved materials you can find at school or at home, including paper clips, pipe cleaners, plastic straws, notecards, tape...

*TEACHER'S NOTES: You can add extra difficulty, like making at least four seats or boxes for army men. Each army man must make at least one rotation/ revolution through the ride for it to be considered a success.*

### QUIZ GRADE:

Research and design on ferris wheels and thrill rides.

- A paragraph on ferris wheels 25%

- A concept idea for your ferris wheel sketched and explained 50%

- Conclusions and reflections based on your results 25%

### TEST GRADE:

Your completed design and the results of the test.

- Project Completed = 50%

- 50% of your grade depends on if your project actually works. More rotations = better grade.

- *NOTE: The best project gets an automatic 100%.*

### NOTES:

**CATEGORIES:** Gears, Rotation, Rubber Bands, Scavengers

## MISSION: Andrew Frinkle

**BRIEF:** A quick look at the author of this book and the previous volume (which I hope you have!).

### ABOUT THE AUTHOR:

1. Over 10 years of experience in the teaching field with a specialization in math and science education in elementary and middle schools.

2. Award Nominated for teacher of the year.

3. Winner of the Karen Pelz Writing Award for short fiction.

4. Author of over 20 books in nonfiction and fiction genres.

5. Black Belt in the Korean Sword Martial Art Geomdo.

### SNAZZY PHOTO:

### CONTACTS DETAILS:

**Email me:**

underspacewar@yahoo.com

**VIsit me:**

- www.underspace.org
- www.littlelearninglabs.com
- www.common-core-assessments.com
- www.veleriondamarke.wordpress.com/

### NOTES:

Chicago style is the best style of pizza.

Chili Cheese Dogs are pretty awesome, too.

**CATEGORIES:** Hands-On, Labs, Math, Measurement, Physics, Science, STEM

## MISSION: SUPER COOL BLANK PAGE!

You always have to publish an even number of pages, so this page has been added for your visual enjoyment! :)

**CATEGORIES:** Make sure you got the original book: 50 STEM LABS

Made in the USA
Columbia, SC
16 February 2018